COMPETITIVE MATHEMATICS 3

INTRODUCTION

This objective mathematics series provides a basic and challenging problem of mathematics from particular topics. It can be used to brush up ones basics and checking up the preparation level of particular topics. It is equally helpful to the traditional classes as well as competitions. It can be also taken as a revision material for any competition which includes the test of basic mathematics. If you want to grasp the subject before practicing these multiple choice questions, you can go through the website http://www.ncert.nic.in/ncerts/textbook/textbook.htm and down load the free copy of mathematics books and after having command on the topic practice it. For revision purpose, important points are given at the starting of each topic.

CONTENTS

10. NUMBER SYSTEM

- A rational number can be written in from of P/U where U≠0 and P and Q are integers.
- A rational number has decimal expansion terminating or non terminating but repeating.
- An irrational number has decimal expansion non terminating non repeating.
- A univalve point on the number line represents every real number.
- The collation of rational and irrational number is read number.
- For two positive real number a and b;
  1. $\sqrt{ab} = \sqrt{a} \times \sqrt{b}$
  2. $\sqrt{\frac{a}{b}} = \frac{\sqrt{a}}{\sqrt{b}}$
  3. $(\sqrt{a} + \sqrt{b})(\sqrt{a} - \sqrt{b})$ =a −b
  4. $(a + \sqrt{b})(a - \sqrt{b})$ =a²−b
  5. $(\sqrt{a} + \sqrt{b})^2$ =a +2$\sqrt{ab}$ + b
- Two set rational is denominate of $\frac{1}{\sqrt{r} + \sqrt{y}}$ we multiph, this by $\frac{\sqrt{x} - \sqrt{y}}{\sqrt{x} - \sqrt{y}}$ where X and Y are integers.
- Low of exponent of real no for (a) $>$ 0

1 $a^0 = 1$

2 $a^m . a^n = a^{m+n}$

3 $(a^m)^n$

4 $a^m / a^n = a^{m-n}$

5 $a^m b^m = (ab)^m$

6 $a^{-m} = 1/a^m$

10. NUMBER SYSTEM

1. Which is a rational number?

   a. 5/0   b. 1   c. $5\sqrt{2}$ d. 1.0010001001

2. Which sequence of number is called is a whole number?

   a. 1,2,3,4,5   b.0,1,2,3,4,5   c. 5,6,7,8,9   d. All of these

3. Which of the following is co – prime?

   a. 7, 11   b. 5, 8   c. 3, 9   d. None of these

4. Which of the following is not an integer?

   a. 3/3   b. 6/3   c. 9/3   d. 5/3

5. Which of the following is natural number?

   a. 8/2   b. 5/3   c. 0   d. 5/2

6. A real number consists?

   a. Rational number   b. Irrational number

   c. Both (a) & (b)   d. None of these

7. Which of the following is an irrational number?

   a. $\sqrt{25}$ b. $\sqrt{16}$ c. $\sqrt{36}$ d. $\sqrt{24}$

8. Write the decimal expansion of 37/99?

   a. 1.3737   b. 0.37   c.0.27   d. 0.17

9. Write 0.83 in the form of p/q?

   a. 5/6   b. 3/6   c. 4/6   d 2/6

10. Which of the following is a rational number?

a. $\sqrt{49}$ b. $\sqrt{35}$ c. $\sqrt{38}$ d. $\sqrt{50}$

11.   Which of the following are currying?

a. 7/8   b. 1/3   c. 6/5   d. ½

12.   Which of the following are terminating?

a. 7/8   b. 1/3   c. 1/7   d. 14/11

13.   If we simply $(5+\sqrt{3})(5 - \sqrt{3})$ we get?

a. 25   b. 5   c. $23\sqrt{5}$ d. 22

14.   If we simply $(\sqrt{7} + \sqrt{5})(\sqrt{7} - \sqrt{5})$ we get?

a. 12   b. 2   c. $7\sqrt{2}$ d.5

15.   If we simply $(\sqrt{10} + \sqrt{5})$ we get?

a. $10+\sqrt{5}$ b. 25   c. $15+2\sqrt{50}$ d. $10+2\sqrt{50}$

16.   If we simply $(2+\sqrt{5})(3 + \sqrt{6})$ we get?

a. $\sqrt{6} + \sqrt{8} + 2\sqrt{5}$ b.$6+2\sqrt{6} + 3\sqrt{5} + \sqrt{30}$

c. $2\sqrt{5} + \sqrt{3} + 18$ d. $2\sqrt{5} + \sqrt{5}$

17.   The decimal expansion of rational number?

a. Terminating   b. Non – terminating

c. Both (a) and (b)   d. None of these

18.   The decimal expansion of irrational numbers is?

a. Terminating   b. Non –terminating

c. Both(a)&(b)   d. None of these

19.   Who was the first t complete digits in decimal expansion of its?

a. Aryabhatta   b. Pythagoras   c. Archimedes   d. G. Cantor

20.   Which is a type of non- terminating and non –recurring decimal expansion?

      a. 0.25   b. 22/7   c. 7/8   d. ½

21.   Which is a type of terminating decimal expansion?

      a. 1/3   b. 233/990   c. $\sqrt{2}$ d. 8/5

22.   If we add, subtract, multiply or divide two rational numbers we get?

      a. Rational numbers   b. Irrational number

      c. Both (a) & (b)   d. None of these

23.   If we add , or subtract between rational and irrational numbers we get?

      a. Rational numbers   b. Irrational numbers

      c. Both (a) & (b)   d. None of these

24.   If we get multiply or divide two rational and irrational numbers we get?

      a. Rational numbers   b. Irrational numbers

      c. Both(a)&(b)   d. None of these

25.   If we add , subtract , multiply or divide two irrational numbers we get?

      a. rational numbers   b. Irrational numbers

      c. Both(a)&(b)   d. None of these

26.   If we solve $3\sqrt{5}^{\frac{\sqrt{5}}{}}$ we get?

      a. 3   b. 4   c. 5   d. 7

27.   If we rationalise the denominator of $5/5+\sqrt{3}$ we get?

a. $5 - \sqrt{3}$　　b. $10 - \sqrt{3}$　　c. $7 - \sqrt{3}$　　d. $4 - \sqrt{3}$

28.　　If we rationalise the denominator of $3/\sqrt{5} - \sqrt{3}$ we get?

a. $3(\sqrt{3} + \sqrt{5})$　b. $3(\sqrt{5} + \sqrt{3})$　　c. $3(\sqrt{5} + \sqrt{3}/2)$　d. $3(\sqrt{2} + \sqrt{3})$

29.　　If we get rationalise the denominator of $5/2 - \sqrt{7}$ we get?

a. $5 + \sqrt{7}$　　b. $-10 + \dfrac{\sqrt{7}}{3}$　　c. $3\dfrac{\sqrt{7}}{3}$　　d. $10 + \dfrac{\sqrt{7}}{3}$

30.　　If we simplify $(5\sqrt{3} + 2\sqrt{2})(5\sqrt{3} - 2\sqrt{3})$ we get

a. $50 - 2\sqrt{2}$　　b. $25 - 2\sqrt{2}$　　c. $75 - 2\sqrt{2}$　　d. $5\sqrt{3} - 2$

31.　　Write the decimal expansion of 5 20/3?

a. 11.6　　b. 11.6....　　c. 11.5　　d. 11.44

32.　　Which type of decimal expansion 12/5 have?

a. Terminating　　　　b. Recurring

c. Non – terminating　　　d. Both (a) & (b)

33.　　If we simply $(4^8 * 5^6)$ we get?

a. 20　　b. 20^{-1}　　c. 20^2　　d. 1

34.　　If we simplify $(5^8/5^3)$ we get?

a. 10^2　　b. 5^5　　c. 5^{11}　　d. 3^5

35.　　If we simplify $[(4^2)^5]$ we get?

a. 5　　b. 4^5　　c. 4^3　　d. 4^{10}

36.　　If we simplify $(18^4 * 8^4)$ we get?

a. 10^4　　b. 144　　c. 144^4　　d. 18^4

37.　　If we simplify $(1/15^3)$ we get?

a. 15^3 b. 15^{-3} c. 15   d. 15^{1+3}

38.   If we simplify $(15^{15} * 15^{-8})$ we get?

        a. 15^{13} b. 15^3 c. 15^{-3} d. 15^{40}

39.   If we simplify $[(5^4)^{-5}]$ we get?

        a. 10   b. 5^9 c. 5   d. 5^{-20}

40.   If we simplify $[5^8/5^6 * 2^6/2^4]$ we get?

        a. 7^{48} b. 10^3 c. 15^{24} d. 10^2

41.   If we simplify $(5^{2/3} * 5^{1/3})$ we get?

        a. 5^2 b. 5^3 c. 5   d. 5^4

42.   If we simplify $(7^{1/3})$?

        a. 35   b. 35/3   c. $7^{5/3}$ d. 7

43.   If we simplify $(5^{1/3}/5^{1/2})$ we get?

        a. 5^2 b. $5^{-1/6}$ c. 5^{-1} d. 5^3

44.   If we simplify $(5^{2/2} * 5^{2/2})$ we get?

        a. 5^5 b. 5^2 c. 5   d. 5^3

45.   If we simplify $[5^{2/5} /5^{2/2} * 8^{2/5}/8^{2/2}]$ we get?

        a. 40^2 b. $40^{-3/5}$ c. 40^5 d. 10^3

46.   If we simplify $(25^{1/2})$ we get?

        a. 10   b. 5   c. $5^{1/2}$ d. 5^2

47.   If we simplify $(\sqrt[3]{2}$ we get?

        a. $2\sqrt{3}$ b. 2   c. $2^{1/3}$ d. 2^3

48.   If we simplify $5\sqrt{25}$ we get?

a. 2.5   b. $5^{1/2}$ c. 25^5 d. $25^{1/5}$

49.   If we simplify ($15^{2/3}$)we get?

a. $25^{1/3}$ b. $15^{1/3}$ c.225^5 d. $20^{1/3}$

50.   If we simply ($20^{2/4}$) we get?

a. 20^2 b. $400^{1/4}$ c. 20^4 d. 20

Answers:

Q	A	Q	A	Q	A	Q	A	Q	A
1	B	11	B	21	D	31	B	41	C
2	D	12	A	22	A	32	D	42	C
3	A	13	D	23	B	33	C	43	B
4	D	14	B	24	B	34	B	44	C
5	A	15	C	25	C	35	D	45	B
6	C	16	B	26	A	36	C	46	B
7	D	17	C	27	B	37	B	47	C
8	B	18	B	28	C	38	C	48	D
9	A	19	C	29	B	39	D	49	C
10	A	20	B	30	C	40	B	50	B

11. REAL NUMBER

SOME IMPORTANT POINTS

➢ Euclid`s division lemma stated as given integer a, b there exist unique integer q and r satisfying. a = bq+r, $0 \leq or\ b$.

➢ Fundamental theorem of arithmetic every composite no. can be factorised as a product of primes, and its factors are unique, a part from the order in which prime factors occurs.

➢ Let q be a prime no. If q divides p^2, the q divides p where p is a positive integer.

➢ Let x be a rational no. whose decimal expansion terminates. Then x can be expressed in form of p/q are co prime, and the prime factorization of q is the form $2^n\ 5^m$ where n, m are non integers.

➢ Let x = p/q be a rational no. such that the prime factorization of q is the form of $2^n\ 5^m$, where n, m are non negative integer. Then x = has decimal expansion which is non terminating repeating.

➢ H.C.F (a,b)*L.C.M(a,b) = a*b.

➢ Sum and difference of rational and irrational number is irrational.

➢ Product of rational and irrational no. is irrational.

11. REAL NUMBER

1. Which of the following number always ends with the digit 1 where n is natural number.

   (a) 11^n

   (b) 4^n

   (c) 6^n

   (d) 8^n

2. Which of the following number is irrational number?

   (a) 3.333........

   (b) 3.01001001001......

   (c) 2.3789

   (d) 9.080706004003....

3. The decimal expansion of the rational numbers $36/4*2^3*5^3$ will terminate after................decimal places.

   (a) 3

   (b) 4

   (c) 5

   (d) 6

4. What is the product of two consecutive natural numbers?

   (a) odd number

   (b) prime number

   (c) an even number

   (d) none of these

5. Which of the following is irrational numbers?

   (a) 2.57

   (b) 3.040040004...

   (c) 3.3333

   (d) none of these

6. Which of the following is a rational number?

   (a) 2.57755775.....

   (b) 2.343443444....

   (c) 7.87654321765....

   (d) both a and c

7.   What is the value of 0.7 to 0.2?

        (a) 0.72   (b) 0.27

        (c) 0.9   (d) 0.79

8.   If in unit digit of 4^3 is 4 what will be the units digit of 4^9.

        (a) 4   (b) 6

        (c) 1   (d) 8

9.   Given that HCF (105,175)=35 find L.C.M (105,175)

        (a) 524   (b) 525

        (c) 624   (d) 1625

10.   Express 100000 as product of prime factors.

        (a) 2^4*5^4 (b) 2^6*5^6

        (c) 5^3*2^3 (d) 2^5*5^5

11.   Which of the following prime number will be repeated multiplied in prime
        factorization of 3600.

        (a) 2, 3, 5   (b) 4, 9, 2

        (c) 2, 5, 12   (d) 2, 9, 18, 12, 5

12.   What is digit at units places of 6^n where n=7

        (a) 4   (b) 6

        (c) 8   (d) 2

13.   Find the digit at unit place of 3^n where n=5

(a) 9

(b) 7

(c) 3

(d) 1

14. Which of the following are prime factors of number or fraction 108/13?

(a) 2^2*3^2

(b) $2^2*3^2*5^1$

(c) $2^2*5^2*3^1$

(d) $5^2*3^2*2^1$

15. If H.C.F of two numbers is 171 and 152 is 19.

(a) 1368

(b) 1868

(c) 1468

(d) 1268

16. What is the H.C.F of 115, 184

(a) 15

(b) 13

(c) 19

(d) 23

17. Which of the following number given quotient and remainder 3 and 5 when divided by16

(a) 83

(b) 55

(c) 53

(d) 85

18. Find the sum of 0.2+0.5

(a) 0.7

(b) 0.8

(c) 0.25

(d) 0.67

19. Which of the following number given remainder 7 and quotient 8 when the number is divided by 151?

(a) 18

(b) 17

(c) 16

(d) 19

20. Which of the following smallest irrational number $\sqrt{32}$ be multiplied by so to get a rational number.

(a) $\sqrt{8}$

(b) $\sqrt{2}$

(c) $\sqrt{4}$

(d) $\sqrt{32}$

21. Find the product of $(\sqrt{9} - \sqrt{6})(\sqrt{9} + \sqrt{6})$

(a) $\sqrt{3}$

(b) $\sqrt{15}$

(c) 3

(d) 15

22. Expressed the number 0.33333...... in rational number

(a) 1/3

(b) 2/3

(c) 3/3

(d) 4/3

23. What is the rational number when 0.262626..... expressed in the form of p/q.

(a) 2/99

(b) 26/99

(c) 23/99

(d) 45/99

24. What is the rational number 0.585858.... is expressed in the form of p/q

(a) 173/99

(b) 27/99

(c) 58/99

(d) none of these

25. Which of the following smallest irrational number should be added to in 9+ $\sqrt{2}$ to set a rational number?

(a) $\sqrt{2}$

(b) $\sqrt{2}$

(c) $9-\sqrt{2}$ (d)$3- \sqrt{2}$

26. Which of the following smallest rational number should be added in $3- \sqrt{2}$ to get a rational number?

(a) 3   (b) $\sqrt{2}$

(c) -3   (d) $- \sqrt{2}$

27. Give the fractional form of 0.25

(a) 25/99   (b) 25/90

(c) 23/0   (d) 23/99

28. Give the fractional form of 2.73

(a) 146/90   (b) 243/99

(c) 231/90   (d) 27/25

29. Give the fractional form 5.23

(a) 518/99   (b) 518/90

(c) 252/99   (d) 123/90

30. Which number should be multiplied to $(\sqrt{3} + \sqrt{7})$ get a rational?

(a) $\sqrt{3} - \sqrt{7}$ (b) $\sqrt{3} + \sqrt{7}$

(c) $\sqrt{7}$ (d) $-\sqrt{13}$

31. Find $3\sqrt{6}+7\sqrt{6}$

(a) $6\sqrt{10}$ (b) 10^6

(c) 60   (d) 6

32. Simplify $\sqrt{3}*\sqrt{5}*\sqrt{21}*\sqrt{35}$

(a) 105   (b) 115

(c) 1325   (d) 17.5

33.   Simplify $\sqrt{28}*\sqrt{8} * \sqrt{14}$

(a) 58   (b) 56

(c) 112   (d) 224

34.   Find the decimal expansion of 7/11

(a) 0.6   (b) 0.3

(c) 0.63   (d) 0.73

35.   Find the decimal representation of 11/9

(a) 1.23   (b) 1.22

(c) 12.2   (d) 2.2

36.   Find the expansion of 17/11

(a) 1.6464--   (b) 1.63

(c) 0.67   (d) 1.54

37.   The decimal representation of 4/7 is 0.571428 then 2/7 is

(a) 0.285714   (b) 0.281754

(c) 0.28145   (d) 10.285174

38.   If 3/7=0.428571..... then 6/7 =?

(a) 0.857124   (b) 0.428715

(c) 0.857142   (d) 0.875421

39.   Find the digital units place of 8^7

(a) 4   (b) 2

(c) 6   (d) 8

40   find the digital units place of 7^7

     (a) 9   (b) 7

     (c) 1   (d) 3

41. What will be unit's digit of 31^{31+}?

     (a) 7   (b) 1

     (c) 4   (d) 9

42. Find the digit at unit place of 6^9

     (a) 6   (b) 2

     (c) 8   (d) 4

43. Simplify $(2-\sqrt{7})+(5+\sqrt{7})+(\sqrt{7}+3)+(-\sqrt{7}+3)$

     (a) 14   (b) 7

     (c) 21   (d) 3

44. Simplify $(\sqrt{5}+5)(6-\sqrt{25})$

     (a) $\sqrt{5}+5$ (b) $6+\sqrt{5}$

     (c) $5-\sqrt{5}$ (d) none of these

45. Multiplicative inverse of $\sqrt{7}-3$ is ....

     (a) $\sqrt{7}-\sqrt{3}$ (b) $3+\sqrt{7}$

     (c) $1/\sqrt{7}$ (d) $7-\sqrt{3}$

46. Multiplicative inverse of $2-\sqrt{6}$

(a) $1/\sqrt{6}$ (b) $1/2-\sqrt{6}$

(c) $2-\sqrt{6}$ (d) $2+\sqrt{6}$

47. Simplify: $(7-\sqrt{5})(\sqrt{5}+7)$

   (a) 2   (b) 44

   (c) 42   (d) 4

48. Simplify: $(\sqrt{5}-\sqrt{3})(\sqrt{5}+\sqrt{3})/(\sqrt{5}+3)+(3-\sqrt{5})$

   (a) 2/3   (b) 3/3

   (c) 1/3   (d) 1

49. express as a rational number $(\sqrt{7}-\sqrt{3})$
   $(\sqrt{7}+\sqrt{3})/(\sqrt{5}-\sqrt{3})(\sqrt{5}+\sqrt{3})$

   (a) 2   (b) 4

   (c) 6   (d) 5

50. The L.C.M of two number 18, 16,144 then what is the H.C.F of these numbers.

   (a) 2   (b) 3

   (c) 4   (d) 9

51. Find the square of $(3+\sqrt{5})$.

   (a) $14+6\sqrt{5}$ (b) $8+6\sqrt{5}$

   (c) $8+5\sqrt{6}$ (d) $6\sqrt{5}+5$

52. Simplify $(2+\sqrt{3})^2(2-\sqrt{3})^2$

   (a) 1   (b) 7

   (c) $1+2\sqrt{3}$ (d) $\sqrt{2}+\sqrt{3}$

53. H.C.F of 84 and 105

    (a) 14   (b) 12

    (c) 15   (d) 21

54. Find the value of $(\sqrt{a} + \sqrt{b})(\sqrt{a} - \sqrt{b})$?

    (a) a+b   (b) a-b

    (c) ab   (d) a/b

55. If H.C.F of two number 198,220 is 22 the find L.C.M of these two number?

    (a) 1987   (b) 2200

    (c) 1980   (d) 1982

56. H.C.F and L.C.M of two numbers are 17 and 340 respectively. If one number is what is other number?

    (a) 68   (b) 102

    (c) 51   (d) 119

57. $\sqrt{3}$ is?

    (a) integer   (b) ratio and no...

    (c) irrational number   (d) add number

ANSWERS:

Q.	A.	Q.	A.	Q.	A.	Q.	A.	Q.	A.
1	A	13	C	25	A	37	A	49	A
2	D	14	B	26	B	38	C	50	A
3	A	15	A	27	C	39	B	51	A
4	C	16	D	28	A	40	D	52	A
5	B	17	C	29	A	41	B	53	D
6	A	18	A	30	A	42	A	54	B
7	C	19	A	31	B	43	B	55	C
8	A	20	B	32	A	44	A	56	A
9	B	21	C	33	B	45	B	57	C
10	D	22	A	34	C	46	D	58	
11	A	23	B	35	B	47	B	59	
12	B	24	C	36	D	48	C	60	

12. TRIGONOMETRY

SOME IMPORTANT POINTS

- In a right angled triangle BAC right angled at A
- Sine A=Perpendicular/hypotenuse
- Cos A= Base/hypotenuse
- Tan A= Perpendicular/base
- Cosec A =hypotenuse/perpendicular
- Sec A= hypotenuse/base
- Cot A= base/Perpendicular
- Cosec A=1/sin A
- Sec A=1/cos A
- Tan A=1/cot A
- Sin A=1/cosec A
- Cos A=1/sec A
- Tan A=sin A/cos A
- Cot A=cos A/sin A
- The value of the trigonometric ratios should be memorized from chapter 0^0, 30^0, 45 60^0, 90^0
- Sin (90^0-A)=cos A
- Sec (90^0-A)=cosec A
- Cos (90^0-A)=sin A
- Cosec (90^0-A)=sec
- tan(90^0-A)=cot A
- cot(90^0-A)=tan A
- $\sin^2 A + \cos^2 A = 1$
- $\sin^2 A = 1 - \cos^2 A$
- $\cos^2 A = 1 - \sin^2 A$

- $\sec^2 A - \tan^2 A = 1$
- $\sec^2 A = 1 + \tan^2 A$
- $\tan^2 A = \sec^2 - 1$
- $\csc^2 A - \cot^2 A = 1$
- $\csc^2 A = 1 + \cot^2 A$
- $\cot^2 A = \csc^2 - 1$

12. TRIGNOMETRY

1. If $sinA = 4/5$ then calculate cosA?

    (a) 9/3   (b) 3/6

    (c) 3/5   (d) 7/8

2. 2cotA=1/4, then calculate tanA?

    (a) 1/8   (b) 2/8

    (c) 8   (d) 2

3. consider ΔPQR, right angled at θ, in which PQ=12 units QR=8 units and $\angle PRQ = \theta$. Determine the value of PQ=?

    DIG

    (a) 8   (b) 10

    (c) 9   (d) $4\sqrt{5}$

4. In ΔOPQ right angled at and OP=8cm

    OQ-PQ=1cm. Determine the values of sinQ?

    (a) 15/73   (b) 16/73

    (c) 17/13   (d) 18/73

5. In ΔABC right angled at B and AB=3cm and BC=4cm. then find cosB?

    (a) 4/5   (b) 5/8

    (c) 3/5   (d) 2/5

6. In $\triangle PQR$ right angled at Q, PQ=3cm and $\angle PRQ$=30⁰ Determine the lengths of the side QR and PR.

(a) $3\sqrt{3}$,5

(b) $2\sqrt{3}$,6

(c) $3\sqrt{3}$,8

(d) $3\sqrt{3}$,6

7. In $\triangle xyz$, right angled at Y, XY=8cm and XZ=16cm the N find$\angle XYZ$.

(a) 30⁰

(b)60⁰

(c) 90⁰

(d) 45⁰

8. If sin(A-B)= $\sqrt{3}/2$, cos(A+B)=0⁰ $< A + B \le 90$o, A $> B$ Find A and B?

(a) A=15⁰,B=75⁰

(b) A=75⁰,B=15⁰

(c) A=60⁰,B=15⁰

(d) A=85⁰,B=20⁰

9. Evaluate:

Sin30⁰ cos60⁰+sin60⁰ cos30⁰

(a) 0

(b) $1/\sqrt{2}$

(c) $\sqrt{3}/2$

(d) 1

10. What is the correct answer of 2 tan45⁰/1+tan²45⁰?

(a) sin60⁰

(b) sin30⁰

(c) sin45⁰

(d) sin90⁰

11. 2sin90/1+sin²90 then find what is the correct value for it?

(a) tan30⁰

(b) tan45⁰

(c) tan0⁰

(d) tan90⁰

12. If tan (A+B)=1 and tan(A-B)=1/$\sqrt{3}$: 0⁰ $< A + B \le 90$o;A $> B$. find A and B?

(a) A=19/8,B=17/8

(b) A=17/8,B=19/8

(c) A=75°/2,B=15°/2   (d) A=15°/2,B=75°/2

13.   If $\sin^2 A = 1/8$ then find $\cos^2 A$?

(a) 6/8   (b) 3/8

(c) 5/8   (d) 7/8

14.   If $\sin^2 A + \cos^2 A = 1$ then find tanA?

(a) $\cos A/\sqrt{1 - sin^2 A}$ (b) $\sin A/\sqrt{1 - sin^2 A}$

(c) $\tan A/\sqrt{1 + sin^2 A}$ (d) none of these

15.   If sinA=5/4 and cosA=4/5 calculate tanA?

(a) 3/4   (b) 4/5

(c) 5/4   (d) 4/4

16.   Find the value of

$\sin^2 60 + \sin^2 30/\cos^2 30 + \cos^2 60$

(a) 2   (b) 5/4

(c) 3   (d) 1

17.   find the value of $\sin 20°/\cos 70°$

(a) 2   (b) 1

(c) 3   (d) -4

18.   Find the value of $\cos 50° - \sin 40°$

(a) 1   (b) 2

() 0   (d) 3

19.   If tan2A =cot (A-30°) then find A.

(a) A=10°   (b) A=20°

(c) A=30⁰

(d) A=40⁰

20. If sec5A=cosec (A-6⁰). Then find the value of A.

(a) 10⁰

(b) 20⁰

(c) 70⁰

(d) 30⁰

21. If cosA=20⁰ then find sinA?

(a) 80⁰

(b) 20⁰

(c) 70⁰

(d) 30⁰

22. choose the correct option $8sec^2A-8tan^2A=$

(a)1

(b) 8

(c) 9

(d) 0

23. $1+sin^2A/1+cosec^2A$ chore the correct option.

(a) cos^2A

(b) tan^2A

(c) sin^2A

(d) cot^2A

24. If x=b $sin\theta$ and y=b $cos\theta$ the n find value of x^2+y^2 is

(a) b

(b) 1

(c) 1/b

(d) b^2

25. If $\theta=90^0$ the find $sin\theta$ $cos\theta$ $-cosec\theta$ $sin\theta$

(a) 0⁰

(b) 1

(c) $\sqrt{2}$

(d) $2\sqrt{2}$

26. If $cos(90^0-\theta)$ $sin\theta=1$ and θ is an a cute angle then find θ.

(a) 0⁰

(b) 45⁰

(c) 90^0 (d) 30^0

27. The value of $(1+\sin\theta)(1-\sin\theta)\sec^2\theta = \dots\dots\dots\dots$

(a) 0   (b) 1

(c) $\cos^2\theta$ (d) $\sin^2\theta$

28. $\cos\theta/\sqrt{1 - \cos^2\theta}$ can be written as

(a) $\tan\theta$ (b) $\sqrt{\sin\theta}$

(c) $\cot\theta$ (d) $\sin\theta/\sqrt{\cos\theta}$

29. $\tan\theta = 1/\sqrt{7}$, then $\cosec^2\theta - \sec^2\theta/\sec^2\theta + \cos^2\theta$

(a) 3/4   (b) 1/2

(c) $\sqrt{2}$ (d) $1/\sqrt{2}$

30. If A and B are a cute angles and cosA=sinB then write the value of A+B.

(a) 45^0 (b) 60^0

(c) 90^0 (d) 80^0

31. $\sec^2 30^0 + \cot^2 45^0 = ?$

(a) 6/3   (b) 2/3

(c) 8/3   (d) 7/3

32. $1 + \cot^2\theta = 2/9$ .then what is the value of θ?

(a) 30^0 (b) 60^0

(c) 45^0 (d) 90^0

33. If $\sin 6\theta = \cos 5\theta$ where 6θ and 5θ are acute angles. Find the value of

(a) $90^0/11$ (b) $11/90^0$

(c) $20^0/2$ (d) $50/3^0$

ANSWERS:

Q.	A.	Q.	A.	Q.	A.	Q.	A.	Q.	A.
1	C	8	B	15	C	22	B	29	A
2	C	9	D	16	D	23	C	30	C
3	D	10	D	17	B	24	D	31	D
4	B	11	B	18	C	25	A	32	B
5	C	12	C	19	D	26	C	33	B
6	D	13	D	20	D	27	B		
7	A	14	B	21	C	28	C		

13. APPLICATION OF TRIGNOMETRY

SOME IMPORTANT POINTS

➢ A line drawn from the eye of an observer to the point of object seen by him/her is called the line of sight.

➢ When we raise our head an angle formed due to the line of sight with horizontal when it is above the horizontal level that angle is called angle of elevation.

➢ When we lower our head to look any object an angle formed due to the line of sight with horizontal when it is below the horizontal level. That angle is called angle of depression.

➢ The values of trigonometric ratios help us to find out the height or length of an object, or distance between two objects.

13. APPLICATIONS OF TRIGONOMETRY

1. A tower stands vertically on the ground, from a point on the ground which is 10m away from the foot of the tower .Then angle of elevation of the top of the tower is found to be 60^0. Find the height of tower?

    a. $15\sqrt{3}$ b. $13\sqrt{3}$ c. $10\sqrt{3}$ d. $12\sqrt{3}$

2. A tower stands vertically from the ground which is 5cm high and angle of elevation is 45^0. Find distance of tower from the ground?

    a. 10   b. 2   c. 3   d. 5

3. An observer 2m tall is 29m away from a chimney. The angle of elevation of the Chimney form her eyes is 45^0.What is the height of the chimney?

    a. 10m   b. 5m   c. 9m   d. 31m

4. From a point on the ground of elevation of the angle of a 18m tall building is 45^0 A flag is hoisted at the top of the building and angle of elevation of the top of the Flag staff from p is 60^0. Find the length of the flagstaff and the distance of the Building from the point p (you may take $\sqrt{3}$ - 1.732)?

    a. $10(\sqrt{3} - 1)$m   b. $11(\sqrt{3} - 1)$m

    c. $18(\sqrt{3} - 1)$m   d. $12(\sqrt{3} - 1)$m

5. The shadow of a tower standing on a level ground is found to be 30m longer When the altitude is 30^0 than when it is 60^0 . Find the height of tower?

    a. $10\sqrt{3}$ b. $12\sqrt{3}$ c. $11\sqrt{3}$ d. $15\sqrt{3}$

6. Find the height of a tower if the angle of elevation of top of tower is 30^0 and Horizontal distance from eye to the foot of the tower is known as 100m?

    a. $100\sqrt{3}$ b. $100/\sqrt{3}$ c. $50/\sqrt{3}$ d. $50\sqrt{3}$

7. A vertical stick 10m long cast a shadow 7cm long. At the same time a, tower Casts a shadow 25m long .Determine the height of the tower?

   a. 35.7   b. 37.5   c. 32.8   d. 39.5

8. From the top of a cliff 50m high, the angles of bottom of a tower observed to be 30^0 and 45^0. Find the height of the tower?

   a. 20.13m   b. 21.13m   c. 22.13m   d. 23.13m

9. The shadow of a tower, standing on level ground, is found to be 45m longer When sun`s altitude is 45^0 than when it was at 60^0. Find the height of the Tower?

   a. $45\sqrt{3}\dfrac{\sqrt{3}}{\sqrt{3}}-1$ b. $45\sqrt{3}\dfrac{\sqrt{3}}{\sqrt{3}}+1$ c. $50\sqrt{3}\dfrac{\sqrt{3}}{\sqrt{3}}-1$ d. $\sqrt{3}-\dfrac{1}{50}$

10. The ratio of the length of a pole and its shadow is $\sqrt{3}:1$ .The angle of elevation Of the sun is?

    a. 90^0 b. 60^0 c. 30^0 d. 45^0

11. A ladder of 10m length touches a wall at height of 5m. The angle θ made by with the horizontal is?

    a. 90^00 b. 160^0 c. 45^0 d. 30^0

ANSWERS:

Q	A	Q	A	Q	A
1	C	5	A	9	A
2	D	6	B	10	B
3	D	7	A	11	D
4	B	8	B		

NOTES

www.ingramcontent.com/pod-product-compliance
Lightning Source LLC
Chambersburg PA
CBHW080623180526
45168CB00007B/3031